NATIONAL GEOGRAPHIC

Ladders

LIVING ON THE COAST

Communities We Live In

Surf's UP

by Debbie Nevins

You stand on a beach. Waves roll toward you. They crash on the shore one after the other. You feel the salty spray on your face. The cold water tickles your feet. You look at the top, or **crest**, of the next big wave. You see two surfers on one surfboard. Then you see that one of the surfers is a dog!

The place where people live affects where they work. It also affects the kinds of homes they live in. It even affects what they do for fun. In coastal communities, most people spend their free time on the beach. Some beaches are even dog-friendly. This means that dogs can be on the beach. They can relax in the sun or play in the water. They can sniff beach smells and catch sticks. And for some dogs in California, they can surf on waves! Some dog **athletes** take part in surfing contests. That's why there are so many wet dogs and cheering people on the beaches of California!

Surf dog Jedi catches a wave on a California beach.

Making a Splash

Each year, more than 1,500 people go to Huntington Beach in California for the Surf City Surf Dog Competition. Dog athletes are called "surFURS." They greet each other with tail wags and sloppy licks. Then they hop on their boards and ride the waves. This unusual sport sure is fun to watch.

Many dogs like to surf. But some dogs want to stay dry. They can bark for their favorite surfers. Or they can be in a dog fashion show or a costume contest. Their owners can shop for dog toys. And it is all for a good cause. Some of the money raised at this event goes to dog rescue **charities**. A charity is a group that helps others. Rescue charities find good homes for dogs that need them. Rescue groups helped many of the surFURS find good homes.

< Check out this pup's colorful life jacket. All athletes must wear safety gear to compete.

This dog's name is Code Four. He's dressed up for the costume contest. He won first place in 2010.

Every sport has its heroes. That includes surfing. Buddy is a Jack Russell terrier. He's a famous dog surfer. He's the first dog to be honored by the Surf Dog Hall of Fame.

This dog wears goggles and a bandanna. The goggles keep salt water out of his eyes. The bandanna just makes him look cool.

Surprising SurFURS

Think surfing dogs are surprising? How about a surfing pig or goat? Just about any pet can be taught to surf as long as it likes the water and can swim. Even some cats can learn to surf if they don't hate the water.

Ready for Action

There's a lot to do before the competition. Owners check their surFURS' floating jackets. The jackets keep the dogs safe in the deep water. Some owners wear floating jackets. They also practice surfing with their dogs.

When the competition starts, owners and their dogs swim out to where the waves are just right. First, the owners help their dogs onto the boards. Then they point them toward shore. When the water rises, they let go and the dogs surf away! Some pups ride all the way to shore. Others jump into the water. But don't worry about them. Their owners are there to help.

The judges give each dog a score. They give points based on how long the dog rides and the size of the wave. Standing up gets five points. Sitting gets three points. Lying on the board gets two points. And dogs can get extra points for doing tricks.

∧ Surfer kitty Nicolasa rides the waves in Peru, South America.

∧ Pisco is an alpaca. Alpacas are relatives of camels. He and his owner surf in South America.

From Leashes to **Longboards**

Some people might wonder how owners train their dogs to surf. Most dogs start by learning to be comfortable on a surfboard.

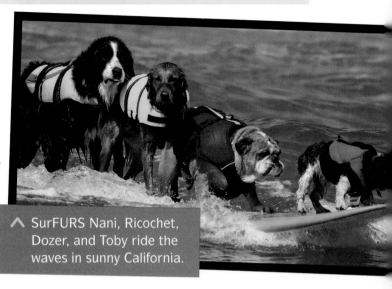

∧ SurFURS Nani, Ricochet, Dozer, and Toby ride the waves in sunny California.

NANI (Akamai Nani Nui)

Nickname: Little Bear

Breed: Bernese mountain dog

Loves: Surfing, swimming, sleeping, and eating

Favorite food: Cheese

Sample honor: Cover girl for the 2010 Surf Dog Calendar

Fun fact: Nani surfed in the movie *Marmaduke* (2009).

RICOCHET

Nickname: Rip Curl Ricki

Breed: Golden retriever

Loves: Surfing with disabled surfers, dock diving, swimming, rolling in stinky stuff, and chasing critters

Favorite food: All foods are her favorites.

Sample honor: 2010 *USA Today* Dog Hero

Fun fact: Ricochet helps disabled children learn to surf.

Some owners bring a board into the house. They give their dog treats for sitting on it. Next, the dog learns to stand on the board. Then it's off to the beach to try some easy surfing. Some dogs learn quickly. Others need more practice. They all enjoy the fun, friendship, and exercise.

DOZER

Nickname: Da Bull 2

Breed: English bulldog

Loves: Surfing, eating, sleeping, dreaming, snoring, and soccer

Favorite food: Everything!

Sample honor: First place, Purina Incredible Dog Challenge Surf Dog Competition, 2011

Fun fact: Dozer appeared in a TV commercial for dog food.

TOBY

Breed: Shih tzu mix

Loves: Surfing, chasing cats and other small critters

Favorite food: Stuffed lamb

Sample honor: First place, small dog category, Surf City Surf Dog Competition in 2010

Fun fact: Toby's owner rescued him as a scruffy stray in an animal shelter.

Check In How do dogs learn to surf?

Read to find out about the coastal communities of the Outer Banks.

Welcome to the

OUTER BANKS

by Jennifer A. Smith

Welcome! The Outer Banks are a group of small islands off the coast of North Carolina. They're called **barrier islands** because they block the coastline from ocean storms. They are separated from the **mainland** by a wide body of water called a **sound**.

Many small coastal communities are part of the Outer Banks. Kids go to school there. Adults work there. And people have a lot of fun living by the water. But it can be tricky to live there during storm season.

> The Outer Banks are more than 175 miles long.

Look at the map. You can see that the Outer Banks are long and thin. They are mostly flat. When the weather is good, the people who live there have fun playing outside. But the barrier islands can get hit by strong storms called hurricanes. These storms can change the coastline of an island. Strong winds and waves can push sand inland. Or they can carry sand out to sea. They can even cut an island in half.

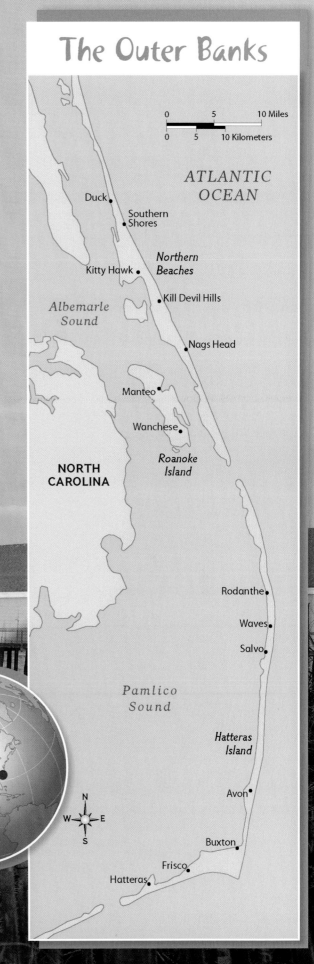

The Outer Banks

0   5   10 Miles

0   5   10 Kilometers

ATLANTIC
OCEAN

Duck
Southern
Shores

Northern
Beaches

Kitty Hawk

Kill Devil Hills

Albemarle
Sound

Nags Head

Manteo

Wanchese

NORTH
CAROLINA

Roanoke
Island

Rodanthe

Waves

Salvo

Pamlico
Sound

Hatteras
Island

N
W E
S

Avon

Buxton

Frisco

Hatteras

First in Flight!

Kitty Hawk, North Carolina, is an Outer Banks community. It has steady winds. It has tall hills of sand, called dunes. In 1900, Orville and Wilbur Wright wanted to test an invention. They called it the Wright glider. A glider does not have an engine. It flies on the wind. Kitty Hawk had good weather for their tests. And it had the right kind of land.

At Kitty Hawk, the Wright brothers could glide off the high dunes. They could land in the soft sand. They later went to another Outer Banks community called Kill Devil Hills. The brothers tested gliders there many times from 1901 to 1903.

> These statues are part of the Wright Brothers National Memorial. It is in Kill Devil Hills.

After a while, the Wrights began working on another
airplane. This plane was powered by an engine. In 1903,
they tried the first powered flight in history. It worked!
Five people watched as each brother took two short flights.
The Wright Brothers National **Memorial** is in Kill Devil Hills.
You can learn more about the Wright brothers' work there.

Fun on the Outer Banks

People who live on the Outer Banks love to have visitors. Tourists can sit on the beach and watch the waves. Or they can do fun activities. Look at some of the fun you can have on the Outer Banks.

> Stunt Kite Competition, Kill Devil Hills

Go fly a kite! A windy beach is a great place to fly a kite. Outer Banks communities hold many kite-flying events. These kites are in the Outer Banks Stunt Kite Competition.

< Lighthouse Tour, Cape Hatteras

Cape Hatteras Lighthouse is the tallest lighthouse in North America. It is more than 250 steps to the top. In this picture the lighthouse looks like it is on the beach. But it is 1,500 feet from the shore.

∧ Hang Gliding, Nags Head

Hang gliders look like large kites. People strapped in hang gliders use the wind to soar over the ocean. You can hang glide at Jockey's Ridge State Park in Nags Head. The park has the tallest natural sand dunes on the East Coast.

∧ Sand Sculpture Festival, Nags Head

Have you ever made a sand castle? You can build castles at this festival in Nags Head. Just grab a shovel and get to work.

Check In What is it like to live on the Outer Banks?

Two Cities, TWO BAYS

by Brett Gover

A cable car climbs a steep hill in San Francisco.

People live in many different coastal communities around the world. Let's compare two of them. Let's see how they are alike and different.

San Francisco, California

Along the Pacific Coast in northern California is San Francisco Bay. Huge ships travel in and out of its deep waters. Ships must pay attention to the flow of water, or **current**, in the bay. These currents are strong. And they change direction often.

Next to the bay is the city of San Francisco. It sits on a **peninsula**, land that is surrounded on

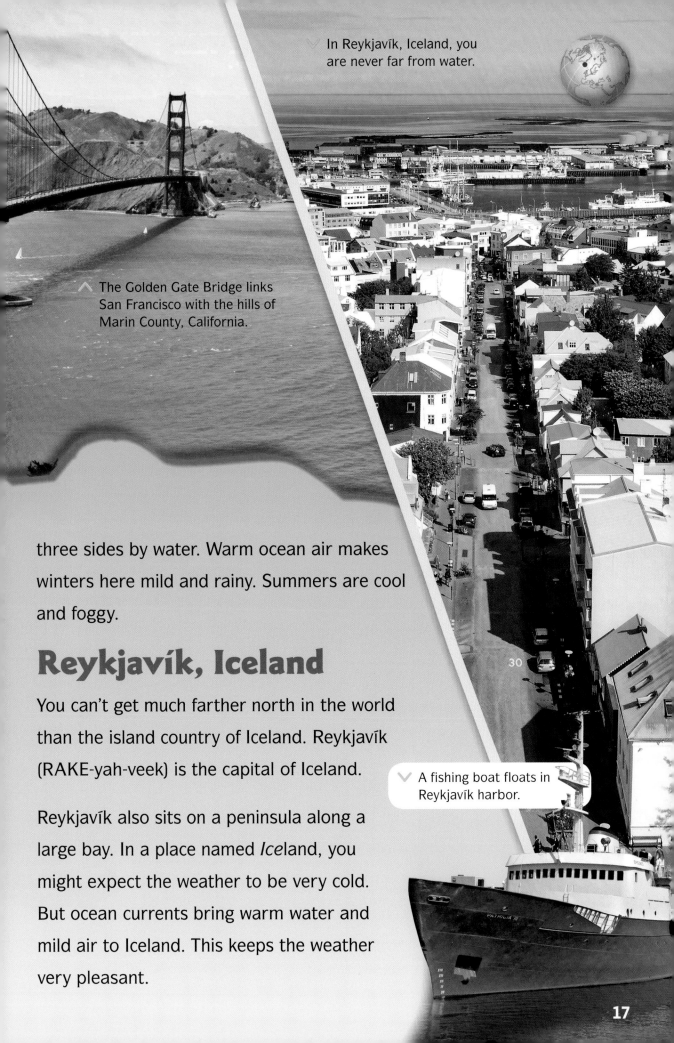

In Reykjavík, Iceland, you are never far from water.

The Golden Gate Bridge links San Francisco with the hills of Marin County, California.

three sides by water. Warm ocean air makes winters here mild and rainy. Summers are cool and foggy.

Reykjavík, Iceland

You can't get much farther north in the world than the island country of Iceland. Reykjavík (RAKE-yah-veek) is the capital of Iceland.

A fishing boat floats in Reykjavík harbor.

Reykjavík also sits on a peninsula along a large bay. In a place named *Ice*land, you might expect the weather to be very cold. But ocean currents bring warm water and mild air to Iceland. This keeps the weather very pleasant.

This boy is panning for gold in the mountains near San Francisco. People have searched for gold here for more than 150 years.

San Francisco Resources

In 1769, Spanish explorers built a small town called Yerba Buena. That town was renamed San Francisco in 1847. Soon after that, many people went there to look for gold.

Have you ever heard of the California Gold Rush? In the 1840s, people found gold in the streams of the California mountains. Many miners moved there. They hoped to find more gold and to get rich. Other people went to open hotels and stores for the miners and their families. San Francisco soon became a busy city.

Reykjavík Resources

People called the *Norse* traveled to Iceland 1,100 years ago. They went to catch fish and to farm. They named their town Reykjavík. This name means "Bay of Smoke." The name came from the steam they saw rising up from the ground. The Norse found that the steam came from hot springs under the town.

Many people of Reykjavík earn money by fishing and raising sheep for **wool**. Wool is sheep's hair. It can be made into blankets and sweaters. Selling wool and fish to other places helped this small village become Iceland's largest and most important city.

Λ Sheep farms are all over the island of Iceland.

Food in San Francisco

Have you ever tried sourdough bread? Gold miners liked its taste. They ate it so much they were called "Sourdoughs." San Francisco bakeries still make the bread every day.

Maybe sourdough bread doesn't sound good to you. You might like something sweeter, like chocolate. In 1849, Domingo Ghirardelli (gir-ahr-DELL-ee) heard about the Gold Rush. He moved to California from Italy. He didn't find any gold. But he didn't leave. He built a chocolate factory. Today, Ghirardelli chocolate is eaten all around the world.

This San Francisco baker shapes the sourdough into loaves. Then he bakes it in the oven. It will turn golden brown.

This fisherman unloads a box of codfish. He is on a ship in Reykjavík harbor.

Food in Reykjavík

If you go to Reykjavík, you will eat a lot of seafood. Fish is very popular there. Fishermen pull salmon, herring, and cod from the ocean all year.

Just as in San Francisco, people in Reykjavík like sweets. Candy shopping there is extra fun on the weekends. Every Saturday, most candy stores sell their treats for half the usual price!

San Francisco Sights

It's easy to have fun in San Francisco. Ride a cable car to the top of Lombard Street. Walk down this steep and curving street. By the time you get to the bottom, you'll be very hungry.

Go to a Chinatown restaurant for a bowl of noodles. You might see a group of dragon dancers on a side street.

Then head to the Golden Gate Bridge for the best view of San Francisco. There is even a path on the bridge for walking or riding a bike.

> Chinatown

⌄ Golden Gate Bridge

∧ Lombard Street

Icelandic volcano

Strokkur Geyser

Blue Lagoon

Reykjavík Sights

Visit the Blue Lagoon near the city. There, you can float in steamy water. If you like sunshine, visit Reykjavík in the summer. At this time of the year, Iceland gets 21 hours of daylight each day.

You can find **geysers** (GY-zurz) all over Iceland. These springs can shoot hot water more than 100 feet in the air. If you think that's cool, you should see a volcano. Near the city of Reykjavík is a volcano that erupted recently. Before that, it had not erupted in about 200 years.

Check In In what ways are San Francisco and Reykjavík alike and different?

Discuss

1. What do you think connects the three selections that you read in this book? What makes you think that?

2. In what ways does the surFURS competition show how important the ocean is to one coastal community?

3. Tell why living in the Outer Banks of North Carolina can be fun, and why living there can be challenging.

4. What effect does a bay have on the lives of the people living in San Francisco and Reykjavik?

5. What do you still wonder about life in a coastal community? How can you learn more?